Astronomy

The Ultimate Beginners Guide To Discover Stars, Galaxies, Wormholes, Black Holes and Astronomy Gadgets

By Joseph Halley

© 2016 Joseph Halley
All Rights Reserved

Table of Contents

Table of Contents

Introduction

Chapter 1: Getting Started Pt. 1 – The Circumploar Constellations

Chapter 2: Getting Started Pt. 2 – Seasonal Asterisms

Chapter 3: Meteor Showers and Planets

Chapter 4: Beyond the Naked Eye

Conclusion

Disclaimer

While all attempts have been made to verify the information provided in this book, the author does not assume any responsibility for errors, omissions, or contrary interpretations of the subject matter contained within. The information provided in this book is for educational and entertainment purposes only. The reader is responsible for his or her own actions and the author does not accept any responsibilities for any liabilities or damages, real or perceived, resulting from the use of this information.

Introduction

The night sky is full of wonders. Most of us spend much of our childhood looking up into the starry sky and wondering just what it is we are seeing up there, transfixed by the beauty of those twinkling points of light. Depending on where you grew up, maybe the sky was filled with stars, so many you couldn't count them all.

Or maybe you could only see the brightest stars, the rest being drowned out by nearby city lights. Maybe you lived (and still live) in a big city where you can't see any at all.

Whatever the case, the stars pull at us with more than mere gravity. There is a certain beauty, a mystical attraction that pulls at us across the vast reaches of space, causing us to look up, night after night.

It has been the same ever since the first man looked up into the sky and thought about what he saw there. Early man could know nothing of what those white specks were made of, giant, superheated balls of hydrogen and helium, but he could still see their transcendent beauty and it sparked his imagination.

As soon as man had language, he filled the night sky with more than stars, he filled it with stories. Stories of hunters, great bears and gods gave new life to the night sky.

Naturally, we do not today look at the sky quite the same way. Animals, men and women are not immortalized by having their souls placed in the heavens. Yet, the constellations imagined by our more poetically inclined ancestors still provide a useful and beautiful gateway into the sky above us. Those bears, kings and queens that populate the sky can be used to find to the brightest, most beautiful stars that inhabit their realm.

This guide will give you the tools necessary to get started navigating your way around the night sky. To be clear, we are talking about the sky as it appears in the Northern Hemisphere, specifically from North America.

Using the ancient constellations as your starting point, you will be able to identify stars, planets and know where to look when there is a meteor shower coming your way.

We'll also point you to some good apps to help increase your knowledge as well as what you should look for in a beginner telescope and how much you can expect to spend.

Now, get yourself a warm blanket, a thermos of coffee or tea and let's get ready to step out under the sky.

Chapter 1: Getting Started Pt. 1 – The Circumploar Constellations

Looking out at the night sky, especially if you live, or have ready access to a rural area without a lot of lights, is truly awe inspiring. The stars can transfix us with their faint rays of light, coming to us from light years away. Yet, it also just seems like a confused jumble of white points of light to the untrained eye.

Most of us can find the Big Dipper and if the sky is sufficiently dark, identify the Milky Way but that is where most people's knowledge of the constellations stops. But if you are reading this, clearly you want to know more.

The best place to begin is with a handful of constellations that are visible throughout the year. These are the Circumpolar Constellations, called such because they circle around the northern part of the sky. These include Ursa Major, Ursa Minor, Draco, Cassiopeia, Auriga, Cepheus, Camelopardalis, Lynx, and Perseus.

We'll begin out journey through night sky with Ursa Major, specifically, its tail, better known as the Big Dipper.

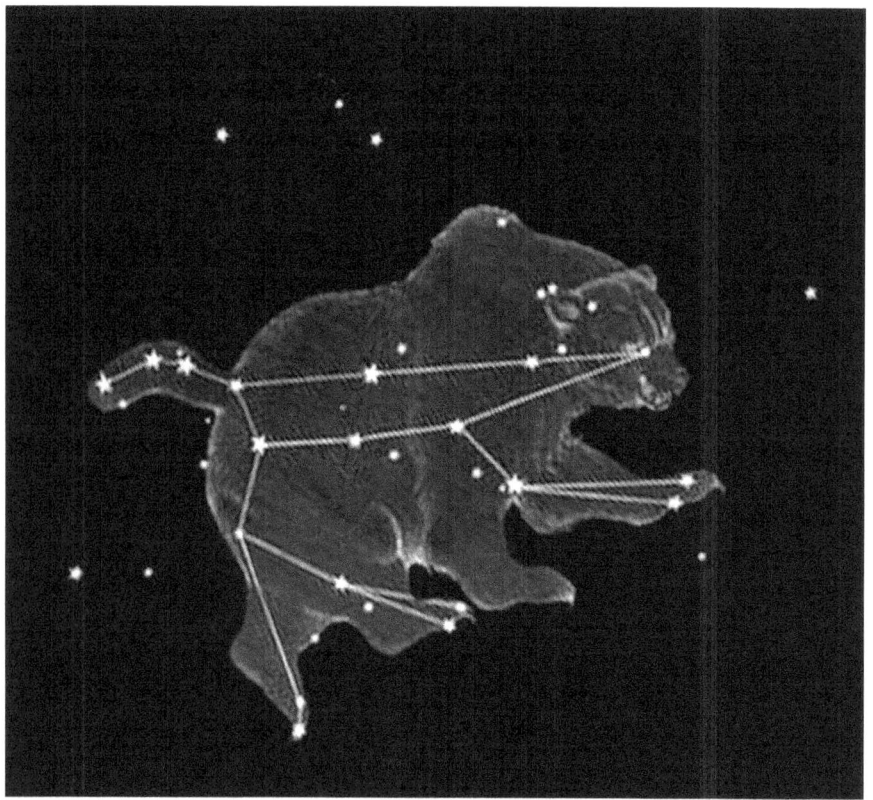

Ursa Major, or the Big Bear, is one of the most recognizable constellations in the Northern Sky. What is more, numerous cultures recognized a bear shape in the sky, each with its own story on how it got there. We'll look at those below.

The bear's tail is the most recognizable part of the constellation. Known as the Big Dipper, the Plow, Wagon and the Wain, this constellation within a constellation is a useful sky navigation tool all by itself.

Taking the two stars (Dubhe and Merak) that form the outside of the bowl and drawing a line out from them takes you to the North Star and the Little Dipper. During the winter months, drawing a line from the handle takes you to the bright star Arcturus in the Constellation Bootes.

A fun fact you can share with your friends is that the second star in the handle, Mizar is actually a double star and it was used by the ancient Greeks and nomadic tribes in the Middle East as a vision test for soldiers and scouts.

 If you could see Mizar's fainter companion, Alcor, you passed the test. It is possible to still see Alcor today even in moderately lit areas.

Mythology: One of the Greek stories concerning the Big Bear is that Zeus got caught having an affair (there are more than a few of those stories) with a young lady named Callisto. Naturally, Hera turned her into a bear and Zeus responded by putting her and her son (Ursa Minor) in the sky to look down from the heavens forever.

The Micmac tribe of Native Americans have an elaborate tale of a bear pursued by hunters, who eventually slay and eat the beast, with its skeleton traversing the sky on its back. In the spring, the slain beast is replaced by a fresh bear from the den and the hunt repeats itself.

Arab and other cultures relate the constellation to a coffin, with elaborate tales of funeral processions describing its movements through the sky.

Ursa Minor (Little Bear) is not only small than Ursa Major, it is also quite a bit fainter. In fact, many beginning star observers will have a hard time locating its most significant star Polaris because it really is not all that significant. The only reason it matters is that it appears to remain fixed in the night sky, with the other stars revolving around it.

It actually does move slightly though and will one day (long after anyone reading this is dead) be replaced by another star. But because of this apparently fixed position in the sky, Polaris has been used as a navigational tool for mariners many times over the centuries.

As a location tool, you can find Draco wrapping itself around the Little Bear with Cepheus and Camelopardalis nearby.

> Mythology: In addition to the myth related above, there is another in which both Ursa Major and Minor saved Zeus from his murderous father Chronos.

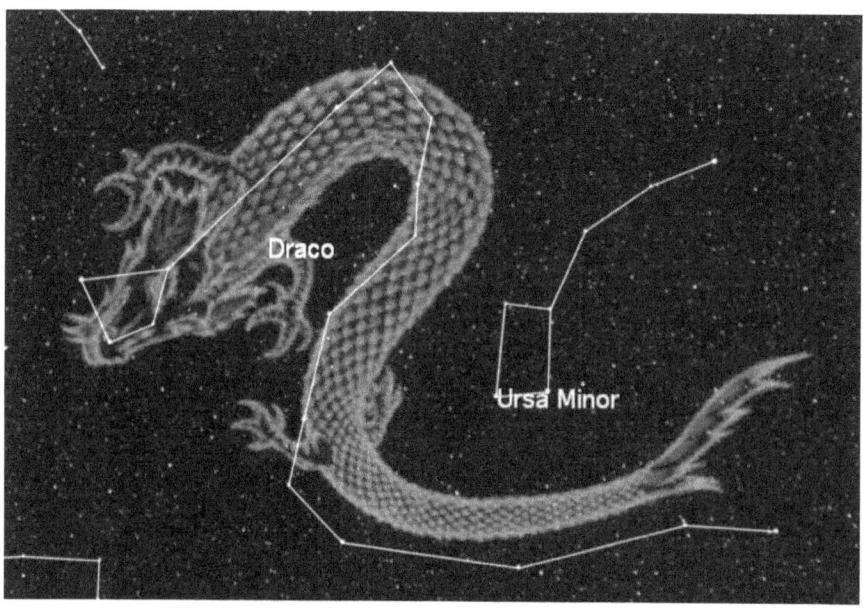

Draco is a long, sprawling constellation, stretching itself out over a significant portion of the sky. As such, it can be difficult to locate if you don't know where to look. Again, if you can find the Little Dipper, you should be able to find the Dragon curving around the bear.

There are also certain stars of note, including former pole star Thuban (way back in 2600 BC) and Eta Draconis and 20 Draconis which are both double stars.

Mythology: Like many constellations, there is more than one myth associated with it. One has Minerva taking Draco by the tale and placing him in the sky during a battle. Another is that Cadmus slayed the dragon while trying to find and rescue his sister Europa.

Arabic astronomy has a completely different view of the constellation. Rather than seeing a dragon, the ancient Arabs so two hyenas attacking a camel which is guarded by four others. Each star represents a different character in this tale.

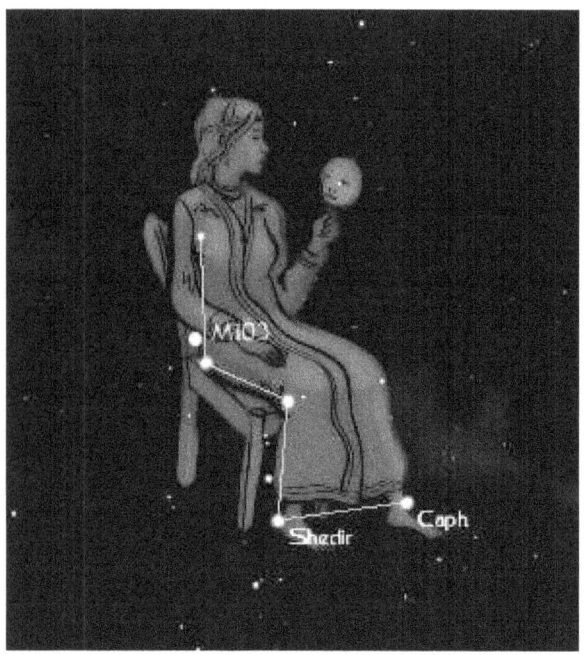

Cassiopeia has a distinctive "M" or "W" shape depending on the time of year, making it one of the easier constellations to locate, though in a dark sky, you may be distracted by the swath of the Milky Way that runs through it.

Using the star on the wider wing of the "W", 45 Cas, one can also find some of the other constellations. Directly above 45 Cas is Errai, which forms the tip of Cepheus. To the left and below slightly is CS Cam, a double star that forms the lower part of the triangle of Camelopardalis. Finally, slightly to the left and farther below 45 Cas is Miran, which makes up the upper part of Perseus.

> Mythology: Cassiopeia was a queen in Greek myth who declared that her beauty was even greater than the legendary sea nymphs. Or, if one prefers the Clash of the Titans version, that her daughter Andromeda was more beautiful than Hera.
>
> Either way, one does not challenge the Greek gods, or even the nymphs and Cassiopeia was punished for her hubris by being placed in the sky for eternity.

Auriga is nestled between Perseus, Gemini and Lynx. It has a shape reminiscent of Cepheus, so attention to nearby constellations is important here. It also has certain stars that are part of the Winter Hexagon, a useful tool for finding many astronomical objects that will be discussed in greater detail below.

Mythology: Auriga was the son of the warrior goddess Athena and is credited with having apparently used the skills inherited from the goddess of battle to invent the four-horse chariot and defeat a rival to the throne of Athens.

Cepheus, thanks to its highly geometric shape – a triangle coupled with a square – is another easy to find constellation. This is fitting since in Greek mythology Cepheus is husband to Cassiopeia, which if you recall, is an easily identified "W" in the sky.

Cepheus also gave its name to the class of stars called Cepheid variables, stars that periodically vary in brightness and have been instrumental in measuring astronomical distances. Delta Cephei was the first variable star discovered back in 1784 by John Goodricke.

> Mythology: As the husband of Cassiopeia and father of Andromeda, Cepheus is of course instrumental in that myth. He is often portrayed with his arms raised, begging the gods to show mercy to his wife and daughter.

Camelopardalis is actually a fairly new constellation, created in 1613 by astronomer Petrus Plancius as a representation of whatever animal it was that Rebecca rode to meet and marry Isaac in the book of Genesis. It was later given its official name by Johannes Hevelius because he thought it resembled a giraffe.

The stars are generally too faint to make the constellation a useful reference point in most areas.

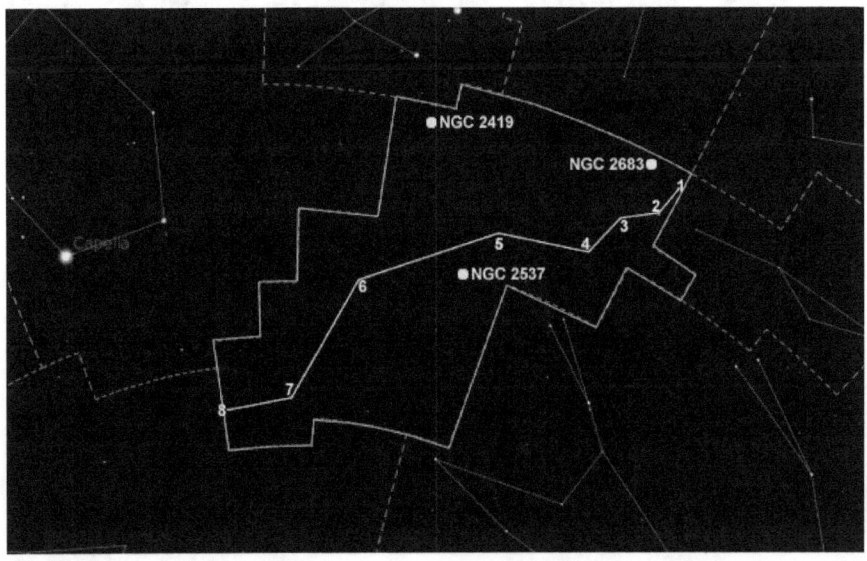

Lynx is also a new constellation introduced by the Johannes Hevelius during the 17th century. Like Camelopardalis above, its stars are very faint, making its utility in the sky minimal at best. However, within its borders is the galaxy NGC 2770 which has had an unusual number of supernovae.

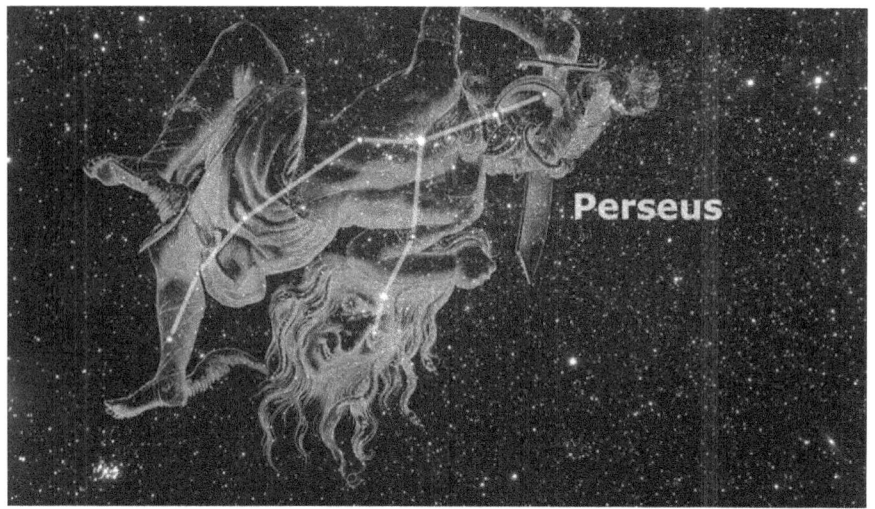

Perseus returns us to the constellations inspired primarily by Greek myth. It is also marked by the Milky Way which passes through the constellation. Its fellow circumpolar constellations Cassiopeia and Andromeda lie nearby to the north and west respectively. Other seasonal constellations Aries and Taurus lie to the south.

Mythology: Perseus was the great hero, and another son of Zeus, who rescued Andromeda from the cliff she had been chained to as punishment for her mother's pride. Before the Kraken could devour her, Perseus showed up, killed the beast and of course rescued the girl.

Chapter 2: Getting Started Pt. 2 – Seasonal Asterisms

Now that we have completed our brief survey of those constellations that are always visible in the North Hemisphere and the utility some have for finding objects in the sky, we will now move onto the seasonal constellations, more specifically, how to locate them and other prominent objects.

We begin with the Winter Hexagon and Winter Triangle. Both are asterisms, which are easily recongnizeable groupings of stars. Technically, any recognizable constellation is an asterism. For our purposes here, we'll be using it refer to a group of stars that is useful for finding one's way around the sky to locate other constellations.

Winter Hexagon

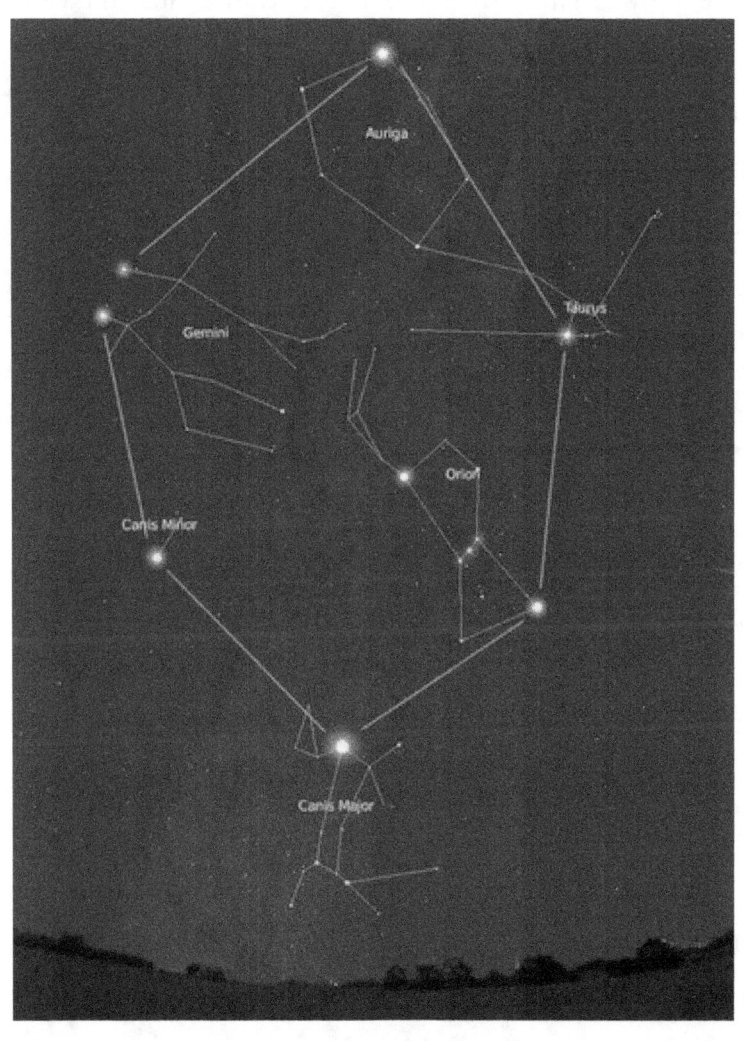

Visible from December to March, the Winter Hexagon consists of the stars Rigel, Capella, Procyon, Sirius, Pollux, and Aldebaran.

- These stars are all part of a constellation which makes it such a useful navigation tool.
- Aldebaran – Right eye of Taurus the bull
- Sirius – Canis Major
- Pollux – Twin brother of Castor, in the Gemini constellation
- Procyon – Canis Minr
- Capella – A double star in Auriga the Charioteer
- Rigel – Lower right of Orion the Hunter

One thing to keep in mind is that this Hexagon is huge. Just one leg is approximately a third of the sky. That means that you aren't just going to look up and spot the shape.

Chances are, you won't be able to see most of it on any given night since most likely clouds will be covering at least one star.

But again, since it is made up of many of the brightest starts in the sky and connects with many constellations, it is still a useful tool for navigating the sky.

Winter Triangle

I actually learned about the Winter Triangle during a presentation by my local astronomy club. I was actually very impressed by how it could be used to locate so many stars and constellations in the winter sky. This was especially promising as I had never been particularly good at identifying constellations before.

The Triangle is actually comfortably nestled within the Hexagon and shares two of the stars, Sirius and Procyon, with the third star being Betelgeuse. It is also very easy to locate as Betelgeuse forms the left shoulder of Orion, probably the best known constellation in the winter sky. Sirius and Procyon are also part of Canis Major and Canis Minor respectively.

Sirius is also the brightest star in the sky, about twice as bright as its closest competition. Betelgeuse will eventually take its place for a little while as it is expected to go supernova in a million years or so. In the meantime, it is the eighth brightest star in the sky.

The fact that the Triangle is so much brighter on average than the Hexagon as a whole and that it is smaller, making it easier to see the whole thing, and is concentrated near the highly recognizable Orion constellation, contribute to making it in many ways a more useful navigation tool than the Hexagon.

Drawing a line almost straight up from Procyon will take you to Capella in Auriga, which also happens to be the second brightest star in the sky.

Up and slightly to the right of Orion's right shoulder (remember his left shoulder is Betelgeuse) will take you to the red-colored Aldebaran in Taurus.

Finally, if you use Betelgeuse and Procyon, you can form another almost equilateral triangle with Castor in the constellation Gemini.

And no discussion of the winter sky would be complete without at least a mention of the Pleiades, or Seven Sisters.

The sisters are an open cluster of seven stars in the constellations Taurus. Depending on your lighting conditions, you may be able to resolve some of the individual stars or they may be merely a bright, fuzzy spot in the sky. But even modest magnification will bring that fuzzy spot to life.

Even a minor tool like a range finder you would use on a golf-course is enough to give you an idea of their beauty. A mid-range set of binoculars that you can pick up for under $30 from the hunting section of any department store lets you see much more.

 There is an almost ethereal beauty about them due to their naturally blue light shining through a gas cloud the cluster is passing through, all of which is visible with that modes $30 investment.

Their mythology also fits their position in the sky very well. The story goes that Orion the mighty hunter began to pursue the Seven Sisters and Zeus, in an attempt to rescue them, turned them into stars and placed them in the heavens.

Undaunted, Orion joined them in the sky where he pursues them to this day. And sure enough, on a cold winter night, it is the Pleiades that rise first, with Orion following them throughout the night.

Spring Asterisms

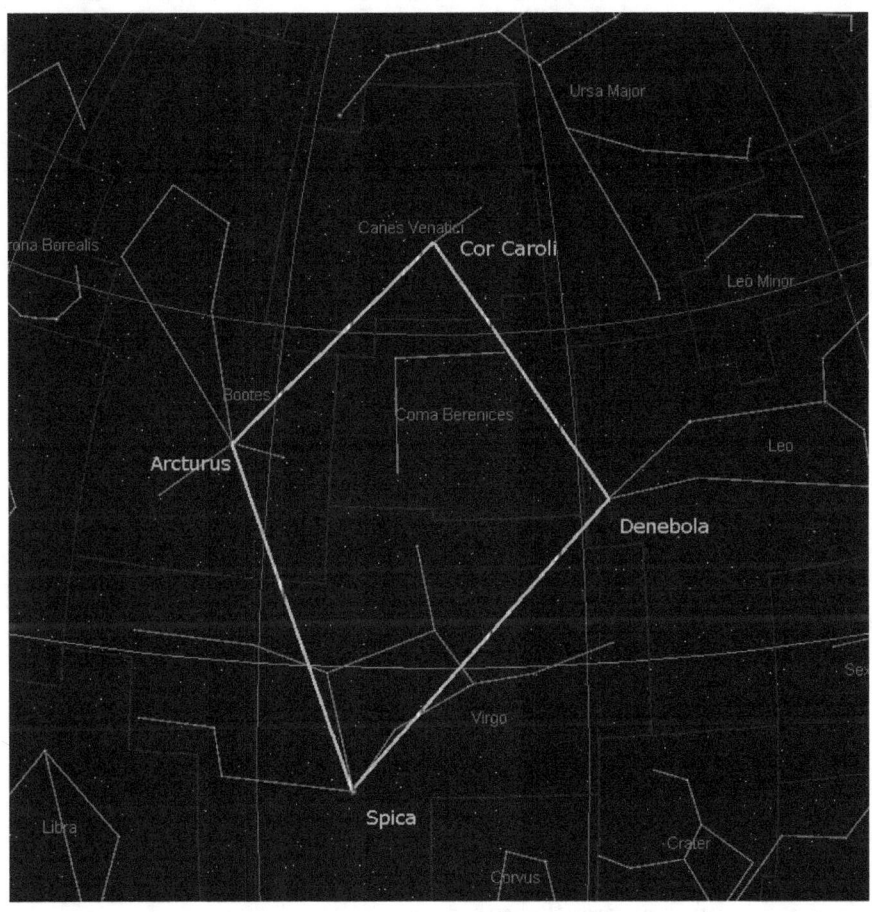

Another fine asterism is the Virgin's (or Great) Diamond, which appears in the spring and is composed of Arturus, Cor Caroli, Denebola, and Spica. This area of the sky does not seem to have much going on. However, if you have a large enough reflector (6 inches or more) then you should be able to resolve many galaxies within the diamond.

Naturally, it also connects to different constellations, with Denebola being the tail of Leo the Lion and Spica being part of the constellation Virgo, Arcturus in Bootes, and Car Caroli (Heart of Charles, named for King Charles I of England) in Canes Venatici.

There are two versions of the Spring Triangle, though both use Arcturus and Spice from the Diamond. One versions makes use of Denebola, another Diamond star, while the other goes farther out to Regulus, which is another star in Leo.

The Summer Sky

In the summer sky, the Big Dipper again comes into play as a useful navigation device. If you draw a line almost straight through the two inner stars of the Dipper's bowl, you will run close to Vega, a star in the constellation Lyra and also one corner of the Summer Triangle.

The Summer Triangle also consists of the star Deneb in the constellation Cygnus and Altair in the constellation Aquila. If you then use Deneb and Vega to form a line back towards the horizon, you come to Antares in Scorpius. I recently was able to recognize this one when a local amateur astronomer pointed it out during an event.

Between Lyra and Scorpius is none other than Hercules himself, while directly down from the star Altair is Sagittarius, the bow and arrow wielding centaur.

Autumn Asterisms

Fall has its own asterism in the Great Square of Pegasus. This asterism is a bit unusual in that it doesn't connect to multiple constellations, it is part of one, namely Pegasus, representing the great winged horse that Perseus rode to rescue Andromeda from the Kraken.

It is however, actually connected with Andromeda, making it handy to help locate the Andromeda galaxy. This the only spiral galaxy that can be seen with the naked eye. Though of course, you can't see the spiral arms as Andromeda is only a fuzzy spot in the sky.

Another nearby object of note is the star 51 Pegasi, which is the first star around which astronomers found an extra solar planet.

Chapter 3: Meteor Showers and Planets

Meteor Showers

A meteor is a bit of an asteroid, comet, or even just a speck of dust that enters the Earth's atmosphere and then burns up due to friction with the air itself, manifesting as a bright streak of light across the sky. If you go out on just about any clear night for at least an hour, you will likely see at least one of these.

A meteor shower, as one might expect, involves considerably more of these bright streaks. This is due to the Earth passing through the debris of an asteroid or comet that passed too close to the sun and was subsequently torn apart. Several of these occur each year and are named for the constellation that they appear to emanate from.

Again, expectations can be an issue here. Calling something a meteor shower many well make you think of a rain shower, with thousands of meteors falling from the sky in a very short period of time.

That, sadly, is not the way it works. Depending on how close the Earth is to the center of the debris, meteors can be seen anywhere from one a minute to one every few seconds during the peak time of the shower.

Unfortunately, the later happens very rarely, with the last truly impressive meteor storm having been the Leonids in 2001. Still, for the patient, there can certainly be rewards. More than one amateur astronomer has acquired some impressive time-lapse photography of a meteor shower.

Below is a list of the meteor showers visible from the Northern Hemisphere, which constellation they radiate from and what season they're visible during.

- **Quandrantids** – Visible in Bootes during mid-winter. A quick point of interest is that this shower gets its name from a now defunct constellation Quadrans Muralis. Like some other constellations mentioned here, it was created relatively recently in the 18th century but for some reason, it never stuck.

- **Lyrids** – Visible in the spring in the constellation Lyra.

- **Eta Aquarids** – Visible a bit later in the spring radiating from Aquarius.

- **Delta Aquarids** – These appear in mid-summer, also from Aquarius.

- **Perseids** – A reliably good show, this meteor shower radiates from Perseus.

- **Draconids** – Radiating from Draco in the fall.

- **Orionids** – Also in the fall, in the constellation Orion.

- **South and North Taurids** – Both of these showers appear in the late fall/early winter in Taurus.

- **Leonids** – Often the same month as the Taurids, the meteors can be found by looking to Leo.

- **Geminids** – These are found in Gemini in the early winter.

The Planets

When our ancient ancestors began to pay attention to the stars, they noticed something odd. Some of the stars seemed to move. Yes, we all know that they do move, but their movement is so slow relative to earth that they appear to stand still from one night to the next.

However, some of these seemed to move around, going forward and backward, up and down in the sky, seemingly with a mind of their own. These came to be known as the planets, or "wanderers."

Overtime, the more observant of our ancestors found that the planets followed predictable paths, making their observation much easier for those of us curious about these wandering points of light even today.

Naturally, one cannot see all of the planets in the solar system with the unaided eye, and those that can are not always visible.

Mercury for instance is too close to the sun. It can be observed but only at very specific times.

Venus is highly visible, being one of the brightest objects in the sky and can be seen in the very early morning, giving this planet the nickname of "The Morning Star."

Mars also is fairly easy to recognize as it has a clear reddish hue to it.

Like Venus, Jupiter is very bright, making it easy to find and will often appear early in the night.

Saturn is much fainter but still very visible.

Some say that it is possible to see Uranus with the naked eye, but as it is at the very edge of what is possible for the human eye to see, most people will never be able to see it without the aid of a telescope. Even if a given person had perfect vision, the viewing conditions would still have to be perfect.

As they each move around quite a bit, from one constellation to the next, it doesn't make much sense to mention all the different places one can look from month to month. Instead, download one of the programs mentioned below that will let you know exactly when each planet is visible and where.

With a good pair of binoculars or a decent telescope (plan on spending a minimum of $200 but preferably $500-600) one becomes able to resolve much more detail.

A modest investment and some practice can bring into view the changes of season of Mars (the polar ice caps change in size), the Galilean moons of Jupiter, the rings of Saturn and even resolve color for Uranus and Neptune. You'll also be able to turn your instrument to other objects in the sky, resolving distant nebulae and galaxies.

Naturally, the more you are willing to spend, the more you will see. That said, don't go all in on a $20,000 "amateur" scope right away. Start small and work your way up. Below, we discuss some of the things to look for in a first telescope.

Chapter 4: Beyond the Naked Eye

As discussed, you can actually see a lot just with the optics you were born with. You can track comets, the movements of planets through the year, gauge the intensity of meteor showers and more. However, if you are interested in astronomy, you naturally want more.

We all want to see the surface of Mars, count craters on the moon, or even discover a comet or two. For those kinds of things, you are going to need either some very high-powered binoculars or a quality telescope.

Now, there is one thing I want to be absolutely clear about. You are not suddenly going to be able to see the deepest mysteries of the universe just because you are gathering a little more light. I remember clearly when I go my first telescope as a child. I was under the impression that I would be able to basically point it at the sky and galaxies and nebulae would essentially leap out of the blackness for my viewing pleasure. If only.

Even if you realize that and understand that you actually have to have objects in mind that you want to look at, it is hard to find them, even if you know where they are.

Your eyes take in a very wide field of view and as such, you can see many things at once. However, as soon as you break out a telescope or a pair of binoculars, you are now significantly narrowing your field of view and that can make it hard to find things.

Not understanding much of this, I became a little disheartened and gave up on my telescope far too quickly.

That is why we spent so much time in the first part of this guide going over how to find the different constellations and some of the brighter stars. The last thing you want is drop $400 dollars (or even a $100) and realize you have no idea what any of that stuff up there in the sky actually is and where the best objects for viewing are.

To that end, there are a couple of very important things to consider when you are buying your first pieces of equipment.

First, consider buying a good pair of binoculars first. You won't be able to see quite as much as you can with a nice reflector telescope but they are easy to put in the car or just take out your front door for a little impromptu moon-gazing. They are also easy to move around as you hunt around for Mars. Or anything else of course.

When you move to a telescope, I strongly recommend getting one that has a motor and a database of objects. You can simply select what you want to view and it will train your scope on the object and even track it as the earth rotates.

Yes, this is cheating a bit, but it is also a good way to learn the sky quickly and get the reward of seeing cool things like the Galilean moons change phase as they orbit Jupiter without going through the more tedious aspects of learning how to adjust a bunch of knobs and levers.

The more you spend on a scope like that, the more objects will be in the database, giving you tons of possibilities for viewing. Other things to consider are the quality of the base and tripod. Most will be made of plastic or aluminum and thus not be terribly sturdy. One small gust of wind can shake the telescope and ruin your viewing until you can get reset. Spend a little money and make sure you get yourself a good steel base that can take being poked and prodded a bit.

Also consider getting one with a camera mount as once you really are sold on this whole astronomy thing, there is a good chance you are going to want to snap a few pictures of your observations. Also be sure to check your manual for directions on how to care for the motor to make sure you get the most out of it.

Apps and Such

Just as with almost anything else, there are number of apps that can help you find your way around the mess of stars over your head, below are just a few of the more popular astronomy apps.

> **Stellarium** – This a free, open source program that lets you view the night sky on your computer, wherever you are. You can enter any location and it will display the stars that should be right over your head.
>
> You can select any object in the sky and zoom in, allowing you to get a close up view of planets and nebulae. The program has many features that make it useful to both the novice and the experienced astronomer.

Star Walk 2 – This app costs under $3 and actually provides an augmented reality experience. Using your GPS, the phone knows where you are and working with your camera, will show you exactly what you see, adding in constellation lines and labels. If you are having a hard time finding the constellations and asterisms discussed here, this is definitely the app for you.

NASA – The official app of the agency that put men on the moon. You can get the latest mission updates, information on stars, planets and past missions. Even better, some versions of the app even allow you to watch NASA TV so you can keep up with everything the agency is doing.

Astronomy Picture of the Day – This is exactly what it sounds like. Every day a new picture of a planet, nebula or distant galaxy is provided, complete with some information so that you know what you are looking at.

Sky Safari – What separates this app from other sky guides is its ability to time travel. Well, not literally, but its software is sufficiently sophisticated that one can use it to see what the sky looked like a million years ago, or what it will look like a million years from now.

Star Chart – If some of those above seem a bit overwhelming, this free app from the Windows Store may be just what you need. It provides easy to follow constellation lines, easy zooming on a given object, and even will play short videos giving you a number of different views of the object selected where available.

Conclusion

The night sky holds many wonders for the patient observer. Learning your way around it has more rewards than just bragging rights. Once you can identify the constellations with ease, you can quickly locate meteor showers when they are supposed to occur and locate the planets throughout the year.

And with a powerful enough telescope, you will be able view the moons of Jupiter, make out mountain ranges on the moon, and even see the rings of Saturn.

While it is true that any astronomy magazine or app will contain far better photos than any you would be likely to take yourself, I can tell you that seeing an up close and personal image of Saturn in a magazine is nothing like seeing them for yourself, even though you can barely distinguish the rings from the planet. Seeing something like that in real time is much better than seeing it in a picture.

And that is the reward for learning the constellations and all the other things we have and will talk about, you get to see it for yourself, without relying solely on the efforts of others. Yes, you are just beginning and you need those others to help you but if you stick with it, you'll soon find that astronomy, like so many things, is well worth the effort.

www.ingramcontent.com/pod-product-compliance
Lightning Source LLC
Chambersburg PA
CBHW070413190526
45169CB00003B/1241